Gulgong
Home rule
Mudgee
Hargraves
Stuart Town
Sofala
Ophir
Orange
Lucknow
Bathurst
Hartley
Carcoar
WALES
SYDNEY
Goulburn
N
0 10 20 30 40 50
SCALE OF MILES

OLD GOLD TOWNS

OF NEW SOUTH WALES

PAINTINGS BY
JOHN DARBYSHIRE
TEXT BY
C. E. SAYERS

RIGBY

STUART TOWN The Main Street

RIGBY LIMITED • ADELAIDE • SYDNEY • MELBOURNE • BRISBANE • PERTH
First published 1971

Library of Congress Catalog Card Number 72-160553
National Library of Australia Registry Card Number and ISBN 0 85179 270 7

Wholly designed and set up in Australia
Printed by Lee Fung Printing Co. Ltd, Hong Kong

OLD GOLD TOWNS OF NEW SOUTH WALES

Even when the "Great South Land" of Australia was known only sketchily, it was predicted that the continent might contain gold and other precious metals. John Harris, in a collection of voyages published in 1704, suggested that gold and silver might be found there because it was in the same latitude as Peru, Chile, and Madagascar, which in those days were famed for their mineral wealth.

The predictions came true in August 1788, when a convict found gold near Sydney. Surgeon John White mentioned the find in his journal of experiences in the first settlement.

From then until 1850 there were frequent reports of gold discoveries, but it seems that official policy was to ignore them. "Put it away, Mr Clarke, or we shall all have our throats cut!" Governor Gipps is supposed to have said to William Branwhite Clarke, the parson-geologist who showed him a specimen of gold in 1844. In a colony with a large convict population, the administration feared a general uprising if such a temptation was displayed.

Government surveyor James McBrien officially reported the discovery of gold in the Fish River, near Bathurst, in 1823, and the Polish explorer-scientist Strzelecki found traces in the Hartley Valley in the Blue Mountains in February 1839: "Sufficient to attest its presence; insufficient to repay its extraction."

Clarke, however, predicted confidently that the country would be found "Abundantly rich in gold." He found it when chipping quartziferous slates at the head of Cox's River, west of Hartley, in February 1841, and made other finds during lonely prospecting trips from the Liverpool Ranges to Bathurst over the next three years. Other finds were made from time to time, but little notice was taken of them. Both Strzelecki and Clarke said later that they had been advised to say nothing about their finds, and no doubt the same advice was given to others.

Official fears were backed up by the general ignorance concerning gold-bearing country and of the simplest methods of winning alluvial gold, but the outlook began to change when the California gold rush of 1849 attracted many men from Australia and had a considerable social and economic impact. As 1850 dawned it was known that payable gold was a certainty in Victoria and New South Wales, but even when the news of the first big finds burst upon Sydney in February 1851 there were still sceptics. Some suspected that the gold washed in Summer Hill creek, near Bathurst, by Edward Hammond Hargraves, had been planted there.

Governor FitzRoy was one of these, and said so in a despatch to Whitehall. But this doubt had not reached London when the first Government geologist, Samuel Sutchbury, reported from the Ophir on 19 May 1851 that, "Gold has been found in considerable quantity . . . I have no doubt of gold being found in greater or less quantity over a vast extent of country."

Hargraves may have been a braggart and his own best profit-seeking publicist, but he had found payable gold. His find was recognised with a reward of £10,000, but the Reverend Clarke had to wait until 1861

HILL END The Old Post Office

before his claim to be the scientific discoverer of gold in New South Wales was recognised publicly with an award of £3,000 by the government. A determined little man with a patriarchal beard, a prospecting pick, and a Bible, he made twenty-eight long and arduous geological surveys, all of them important in the early development of goldmining in Australia.

Hargraves' discovery of gold vanquished the sceptics, and ushered in the "roaring days" of New South Wales. The social and economic effect of the discoveries was enormous. In nine years, the population of New South Wales was nearly doubled. Its import trade increased sevenfold and export trade fivefold, and the influx of goldseekers changed the whole pattern of colonial life. Centres west of the Blue Mountains, pioneered a generation earlier, expanded rapidly. Farmers followed the goldseekers—or they themselves turned into farmers—and opened up much new country. On plains and in river valleys, men laid the foundations of new towns that were to become prosperous centres.

Their growth continues, though it is many years since they relied on gold for their prosperity. Much of their glamour departed as soon as the gold petered out, and even some monuments of the gold years were demolished. But in recent times there has been a revival of interest in the "roaring days," and efforts are being made to preserve what is left of many of the old goldmining centres which flourished in New South Wales.

In some of them, little can be found; perhaps a few old buildings and the remains of mines and mining equipment. Others have more to show, and a number of centres have local museums which cherish remnants of the past. Fortunately, the romantic colour and physical appearance of once busy townships, and the character of the gold diggers and their manner of life, are preserved in a collection of vivid photographs made during the 1870s by Beaufoy Merlin and Charles Bayliss. They were financed by Bernard Otto Holtermann, a German who toured the world with their work. The collection was forgotten after his death in 1885, but rediscovered in 1951 by a grandson in a backyard shed of a North Sydney property once owned by Holtermann. The unique record of the "roaring days" was presented to the Mitchell Library, Sydney.

Artists have done much to revive interest in several of the old gold towns of the central west of New South Wales. Recently, they have re-created much of the exciting atmosphere that Henry Lawson captured in his prose and verse, and their work complements the activity of local historical societies, folk museum promoters, and the government and civic organisations which are striving to preserve and restore what is left of the old gold towns.

Whatever can still be seen of them is history in its most fascinating form; the actual and tangible evidence of the hopes and endeavours of our forefathers. All the enthusiastic effort which is being devoted to their preservation merits the fullest support, because they stand as memorials to one of the most important eras in Australian history and have national, architectural, aesthetic and recreational values which are beyond calculation.

ADELONG White Cottage

Those who seek gold in Tumut nowadays will find it in the autumn glory of its elms, poplars, and other exotics. Together with the picturesque townships of Adelong and Batlow, it attracts many visitors because of the grand scenery in which it stands and because it is the starting point for seeing the mighty Snowy Mountains hydro-electric scheme.

All three towns lie in the district of Tumut in southern New South Wales. Tumut, the name for a river and a town, was the least noted for its gold, though its streams were panned from early goldrush days. Adelong and Batlow have a history of gold finding and exploitation in the middle years of last century.

Diggers laboured through deep snow drifts and thick forests, packing in supplies and equipment on their backs for most of the rugged way, and found gold on Adelong Creek in the mid-1850s. The creek was rushed in 1855 and again in 1860, the peak year, when 28,000 ounces were won by 5,000 diggers working in the heavily wooded valleys and along the icy streams. There was a brief revival in the 1890s, and 18,690 ounces were washed in 1897. Nowadays, the wealth of the district comes from beef, wool, timber, fruit, and vegetables.

Some of the produce is processed in a cannery at Batlow, a small town that tourist literature says, "Nestles in glorious orchard country in the foothills of the Australian alps." First known as Reedy Creek, Batlow was for a time the centre of shallow alluvial diggings.

But gold has long since ceased to glitter in the lovely Tumut district. For half-a-century its fruit has been packed and marketed by a co-operative society, and its trout streams and scenery attract more visitors each year than the thousands of diggers who sought for gold during its early years.

TUMUT Roadside Cottage

BATLOW Old Shed

FORBES Iron Lace in Rankin Street

Soon after the rush along the Lachlan River began, a journalist wrote that the diggings "presented the appearance of a Crusader's camp: tents, huts and hovels of every description that human ingenuity could devise stretched away for miles, and a few good buildings were lost in the general class."

The first gold on the banks of the Lachlan was found in June 1861, by a prospector known only as German Harry. He was prospecting on what was then known as Cunumbla sheep run, which embraced the district of which Forbes is now the centre, and found alluvial gold in such quantity that the news brought in a rush of 8,000 diggers by December 1861. Six months later, an estimated 28,000 diggers toiled along the Lachlan, with the lucky ones working the deep leads that spread out from Forbes township. They won 300,000 ounces in two years, but by the end of 1863 the gold had petered out and two-thirds of the polyglot throng had departed.

The site of Forbes was first seen in 1815, by the explorer-surveyor George Evans. Two years later it was examined by John Oxley, who called it Camp Hill; the name still borne by the town's main residential area.

The town itself was named after Sir Francis Forbes, Chief Justice of New South Wales, and gained a more permanent wealth from its position in the centre of a rich pastoral district. Ben Hall, the bushranger, was shot at a spot five miles outside the town which is now marked by an obelisk, and was buried in Forbes cemetery. But a more elegant reminder of Forbes' colonial days is the group of public buildings which give an almost classical appearance to the centre of the town.

HARTLEY View of Hartley

Once a way had been found through the Blue Mountains, travellers and stock drovers journeyed westwards along a primitive road that descended Mount York by a steep and rugged pass. It passed close to the site of Hartley village, and ran on to the Bathurst plains through the lovely valley which Governor Lachlan Macquarie named the Vale of Clwydd. Now, it is known as Hartley Valley.

A new road to Bathurst was built and maintained by convict gangs, following the easier Victoria Pass route surveyed and planned by Major Mitchell. It was to become the route along which numberless diggers and their camp followers surged to try their luck in the rushes of the central west, passing through the little village of Hartley which had been gazetted in January 1838. It had become an important stopping place for settlers going to the western plains, and possessed a police post and court house which today is a national monument.

Paul Edmund de Strzelecki, the self-styled Polish count who was a pioneer of Australian exploration, found a specimen of gold near Hartley in 1839. He was not to see the village in its heyday. "Romantic, clean, and English-like," it was clamorous with demands for labour of every kind to satisfy the hurrying diggers and their needs; shoemakers, blacksmiths, carpenters, wheelwrights, masons, shepherds, farm labourers. The inn by the river, with its "wax candles, clean linen, good feeding, and an attentive ostler," did a roaring trade.

The court house was kept busy, too, and a police magistrate had to be appointed. This sandstone building, one of the few original structures remaining from the attractive village, is the oldest in New South Wales outside Sydney and has been classified by the National Trust as "a building worthy of preservation."

CARCOAR Stoke House

CARCOAR Old Court House

CARCOAR The Red Brick Shop and Dwelling

As a gold mining centre, Carcoar came comparatively late into the field. Copper, high-grade iron ore, and some gold were found in the district in the mid-1840s, and in 1848 an iron mine was opened on Coombing Park, a noted pastoral property dating from 1825, and worked for some years. But the gold was not fully exploited until 1871, in which year the yield was 20,000 ounces. A rich enough promise, but the diggers succumbed to the lure of rich strikes at Parkes, Gulgong, and Hill End. The Carcoar fields were prematurely abandoned, but saw another brief spurt of activity early in the 1880s—when the "roaring days" were all but gone.

Standing on the central tablelands, Carcoar is a township of great charm. The third oldest town west of the Blue Mountains, it has attractive streets which were designed under town planning regulations laid down when Sir Ralph Darling was governor. The main street, Belubula, is named after the river which flows through the picturesque township.

It is a proud old place, rich in public and private buildings of the colonial era. These include St Paul's Church of England, dating from 1845 and classified A by the National Trust; the stone Roman Catholic Church of the Immaculate Conception, in which the foundation stone is located high up in the north wall; the brick Presbyterian Church, built in 1860; and the original Commercial Bank, built in 1862, which Ben Hall and his gang held up in July 1863.

The court house, opened in 1882, retains its original fittings and has much furnishing which is authentic to its period. The building is being restored by the National Trust, which does all it can to preserve the historic and aesthetic character of Carcoar.

HOME RULE The Little Brick Schoolhouse

MUDGEE The Bakery

Mudgee was a quiet pastoral area when the finding of the great gold mass of Kerr's Hundredweight made every man dream of instant wealth, and caused the rush into the central west goldfields. The small hamlet of Mudgee was a natural centre for goldseekers fanning out through the hills, creeks, and gullies, and grew rapidly as its stores provided tools and sustenance and its grogshops supplied celebration or consolation. To begin with, gold finds in the immediate area were scanty, except for a strike along Lawson's Creek, and most prospectors sought further afield. They might have done better to dig nearby, because rich veins were found in the 1860s at Home Rule and ten years later at Gulgong. Home Rule is a veritable ghost town now, with only a few buildings still standing from those of the booming township of a century ago.

Henry Lawson spent some of his hard boyhood years at Eurunderee, between Mudgee and Gulgong. It is a scrubby area of poor land, but once gave a brief yield of good gold. The site of the cottage where he lived with his parents is now a national reserve.

Mudgee itself stands in a rich upland valley of the Cudgegong River, a tributary of the Macquarie. It is the second oldest town in New South Wales west of the Blue Mountains, and William Lawson, who crossed the mountains with Gregory Blaxland and William Charles Wentworth in 1813, is credited with being the first white man to stand on its site, in 1821. A memorial to him stands in Mudgee's Lawson Park.

George and Henry Cox were the district's first settlers, in February 1821. Others soon followed, and laid the foundations of a great pastoral industry which has since diversified. The red soil slopes of the valley now produce lucerne, wheat, fruit, grapes, and vegetables around the attractive town of Mudgee.

GULGONG The Four Ways

GULGONG The Lace Verandah

GULGONG Herbert Street

For ten years, from 1870 to 1880, Gulgong was the hub of the world for the thousands of diggers and their followers who crowded the roaring township of calico tents, stringy bark and slab huts, hotels, theatres, banks, and two streets of bakers, butchers, chemists, doctors, drapers, and grocers. The tongues of many nations were heard along the crowded streets, and the name of Gulgong won world renown for its imperial bounty of thirteen tons of gold in five years.

The ground was so rich with gold that sparkling particles of it clung to the roots of grass dug up by Tom Saunders, a shepherd. In two hours he had won fourteen ounces of gold; the first real strike at Gulgong, and the trigger for one of the colony's biggest rushes. Saunders struck gold in April 1870, and by June there were 5,000 diggers on the field. Many found gold only just below the surface.

Saunders had worked for the two sons of Richard Rouse, a free settler who arrived in Sydney in 1801. In the early 1820s they took up land along the remote Cudgegong River and developed two sheep stations, Biraganbil and Guntawang, which became fine properties despite harsh travel conditions, comparatively poor land, and Aboriginal hostility.

Guntawang became Gulgong; a village around a church, store, and school. Here, in 1851, the government geologist Samuel Sutchbury picked up nuggets, but he was ordered to Bathurst before any prospecting could be done. Fossickers arrived in 1867, and a gold field was proclaimed in 1868, but Saunders' discovery was the first notable find.

Now, little remains of the "roaring days," though a few buildings are preserved as memorials to the brief glamour of Gulgong.

LUCKNOW Old Mine Shaft

The village of Orange, ten miles from Ophir, became a gold mining centre which grew in importance as rushes were made to diggings in the central west of New South Wales. One of these diggings was named Lucknow, after the great siege in which British soldiers held off Indian attackers during the Indian Mutiny of 1854.

Some gold had been found there, near the main road to Bathurst, in 1851. But it was difficult ground to work, complicated by basalt leads which were unlike most of the ore bodies worked in Australia. In 1862, English capital was brought in to finance organised mining, and though the yields were profitable they were costly to produce. From 1881 to 1900 the more important veins were worked by a London-based company with a capital of £500,000; an operation which produced gold valued at £800,000.

The Orange mining division produced a total of half-a-million ounces of gold, valued at about £3,500,000. Most of this yield came from the mines around Lucknow.

SOFALA The Tall House at the End of the Street

The first gold at Sofala was found after the strike at Ophir, and it became one of the three important diggings of the Turon field. Its heyday population was an estimated 30,000 diggers and 10,000 Chinese, catered for by thirty hotels and grog shanties. Ashton's Circus gave light relief to men tired of toil in creekbeds and mine shafts, and played day and night sessions for six months.

Some big nuggets were uncovered at Sofala. A man named Crosswell found one protruding from the ground inside a cave where he was sheltering from a storm. A party of Chinese, working a claim on Spring Creek, found a mass of gold so large that it took six men to carry it, but its true size was never revealed. They broke it into small pieces and are supposed to have shipped them to China.

Another Spring Creek story is of two diggers who found a 100-ounce nugget. One of them grabbed it and headed for the hills, with his mate in close pursuit and calling plaintively, "Honesty's the best policy!" Honesty won the race—perhaps because the pursuer did not have to carry a handicap of more than six pounds of gold.

The remains of races cut to carry water for washing the pay-dirt can still be seen. One race was seven and a half miles long; another, carrying water for one and three-quarter miles to a claim west of Sofala, produced 300 ounces of gold from its first wash.

Digging the races was a skill known to very few. The Chinese, who worked the field in large numbers, used a cupful of water balanced on a stick for a spirit level when cutting a race.

BATHURST Terrace Row — Piper Street

SOFALA Old Shacks

BATHURST The Old Railway Hotel

Bathurst, the first inland town in Australia, was established in 1815. Most of the explorations into the western and south-western areas started from there, and it was the most important town west of the Blue Mountains when the first payable gold was found at Ophir, only thirty miles away.

It has been said that this find "converted Bathurst from a sleepy provincial town into a prosperous and important centre."

The town developed rapidly as goldmining became the major industry of the central west. Most things that the diggers needed had to be shipped through Bathurst. Whether the diggers won or lost, the Bathurst merchants were able to profit by the food, liquor, tools, clothing, and other supplies which they sold to them, and the Bathurst hotels and other establishments thrived accordingly.

This steady flow of wealth consolidated the city, and as a legacy has given us one of Australia's most attractive communities, with pride in its long history and a sensitive regard for its preservation. Sheep and cattle raising, mainstays of the Bathurst economy before gold was ever dreamed of, remain as valuable industries together with fruit and vegetable growing, dairying, and bee-keeping.

HARGRAVES The Iron House
(by kind permission of Mr. Peter Damman)

STUART TOWN Shady Verandahs

HARGRAVES The Little White Township of Hargraves

The village of Hargraves was first known as Summer Hill, but was re-named to honour the gold discoveries of Edward Hammond Hargraves. It was part of the original Turon River diggings, found by the government geologist, Samuel Sutchbury, in June 1851.

Hargraves, as an emigrant English youth, had battled to make a living with cattle in this hungry country in the early 1830s. He returned to it in January 1851 after a spell on the California gold fields, because the similarity of the two places had made him confident that he would find gold there.

Even though he was not the first to sight gold in New South Wales, his California experiences enabled him to demonstrate how it could be won from the gravels of rivers and creeks. His knowledge caused the name of Australia to flame throughout the world, and its nationhood to be born.

The Hargraves field paid steadily. The uncovering of Kerr's Hundredweight caused a rush that found rich alluvial deposits, and for a long time the gold escort guarded upwards of 5,000 ounces of gold being sent out once a fortnight.

Even after most of the easily won gold had been mined, and most of the diggers had drifted to other rushes, good returns were still obtained. In 1858, 40,685 ounces of gold were won, and from that year until 1874 the total recorded yield was 205,585 ounces—an average yearly yield of 12,000 ounces for seventeen years.

GUNDAGAI The Three Red Chimneys

GUNDAGAI The Hide and Skin Store

For most people, Gundagai's fame is based upon the dog which sat (and still sits, on a memorial) on the tucker box. It is a folklore reminder of Gundagai's long story. Explorer Charles Sturt sailed past its site on his 1829 voyage down the Murrumbidgee; sheepmen and cattlemen rode past into the riverine, highlands and plains; bushrangers marauded nearby; and diggers rushed the place in the early 1860s.

After small strikes in 1858 and 1859, the first large find was in 1861 within the town boundaries. Gundagai boomed for the next four years, to a peak year in 1865. It was never a rich goldfield, though from 1858 to 1875 the recorded production was 46,000 ounces. While diggers came, toiled, and departed, the town grew more solidly upon the district's production of cereals, potatoes, butter, cattle, and sheep.

The gold attracted many notorious bushrangers to the busy main road running south to the Murray. Frank Gardiner robbed the Gundagai mail coach in 1860, and lesser rogues did the same. Ben Hall and two mates bailed it up in 1864, and after a gun battle made off with a rich haul, leaving a dying police trooper behind.

There was a second gold boom in 1894, and for the next few years several mines were worked in the district. The richest was the Prince of Wales at Reno; five miles north of Gundagai. Between 1897 and 1901, its owners produced 23,342 ounces from 72,464 tons of ore. Tribute parties won a further 3,000 ounces up until 1910, but since then the gold of Gundagai has been little more than a memory.

WYALONG At the Crossroads

WYALONG Main Road Shops

"These desolate areas will never again be visited by civilised men," wrote Surveyor-General John Oxley after exploring the Wyalong district in 1817. Some of those who ignored his prediction had cause to regret it. Hardy stockdrovers entered the flat, scrubby Lachlan River plains as early as 1833, but suffered years of crippling drought. A few struggled on, and by the late 1870s most of the country along the Lachlan was occupied by sheep runs.

The Neeld family went there in the 1880s, to grow wheat in the mallee country of the Bland district. Their first house was a bark hut, and they would have remained poor selectors if Joseph Neeld had not seen a piece of stone which glittered dully. He knew it to be gold.

That was in August 1893. Between then and December, Joseph Neeld and his three brothers and their father Frederick applied for a mining tenement. They opened one of the great goldfields of western New South Wales, and their Number One claim was a bonanza. At its height of production, they refused an offer of £250,000 from an English syndicate and still made a fortune. But they received no reward for finding the Wyalong field, even though they had located most of the reefs which attracted 5,000 diggers. When George Neeld sought appraisement of compensation, he was awarded £115.4.0 for loss of improvements and £800 for loss of his selection.

Payable gold was won in early 1894, and the first crushing yielded 571 ounces of gold from 181 tons of stone in early March 1895. Neeld's selection became a village, and the mallee was dotted with tents and crude huts run up by men streaming out from the railhead at Temora. Many of them had had their fares paid by the government, because it was a time of depression and great unemployment.
A newspaper commented: "The fever for fossicking for gold is draining our cities and big towns of the unemployed. Our useless parliament would have seen them starve, but nature has stepped in and the long looked-for good times are coming to pass."

YOUNG Hotel Young

Lambing Flat is a strangely meek name for one of the wildest of the New South Wales gold fields. It stands near Young, an important town on Burragong Creek on the south-western slopes of the State. In the 1830s, James White took up a sheep run which he named Burragong, and established a lambing place on a small plain. When Denis Regan found gold there in 1860, the township that sprang up was named Lambing Flat.

An important find was made there in July 1861, and 15,000 diggers flooded the field. Within a few months, they had won about 125,000 ounces from alluvial leads, but it took a tough man to survive. To begin with there was no police surveillance, and violent crime, gambling, grog selling and standover men went uncontrolled.

Even worse was the demonstration organised by a vigilante committee which had been formed to keep order. They concentrated against the mass of Chinese who by the end of 1860 were crammed into miserable shacks along the creeks. On 12 December a procession raged through the settlement, tore down gambling and grog shanties, then turned against the Chinese and chased them from the diggings. Two were killed.

Thus began the infamous anti-Chinese riots. During January 1861 a Miners' Protection League demanded expulsion of Chinese, repeal of the gold duty, unlocking of public lands, and protection for native industries. Premier Charles Cowper promised consideration if the riots ceased, and armed troopers kept order from March until late May, when dissatisfied diggers demonstrated again. On 30 June, the Chinese dwellings were plundered and burnt, and the Chinese driven from the gold field. Rioting continued, the Riot Act was read on 14 July, shots were fired, and one man killed and several wounded. Soldiers and sailors had to be sent to Lambing Flat to overawe the unruly diggers, but in November 1861 a law restricting Chinese entry to New South Wales was passed—a direct result of the riots.

The town was named Young after Governor Sir John Young.

STUART TOWN Morning Light

The few picturesque remains to be found at Stuart Town recall the days when it was known as Ironbarks. In the 1870s, alluvial and reef mining in the district brought good returns to thousands of men who streamed into the Wellington Valley.

It has been estimated that, between 1875 and 1914, 140,000 ounces of gold were produced at Stuart Town and at Bodangora, a few miles north-east of Wellington. The Wellington Valley fields may have been latecomers into the "roaring days" because diggers were pre-occupied with the great discoveries in the central west. Gold had been seen near Stuart Town as early as 1839, but it was not until other diggings were exhausted that goldseekers turned their eager eyes in that direction.

A. B. (Banjo) Paterson immortalised Stuart Town in his poem *The Man from Ironbark,* and the town still chuckles over that piece of folklore. Nor has it overlooked the golden years. Many of the people who holiday in the district now fossick in the creeks for gold and gemstones, and are given expert advice on gold-washing by the tourist centre of Stuart Town, one of the most attractive of the old towns which was founded on gold.

HILL END Old General Store

It has been said that the fascination of Hill End lies neither in its past nor its present, but in some mystical quality which combines the two. It is a quality which surely must derive from its gold digging past, its recent discovery by artists, and its great charm for visitors.

Few of the buildings which once gave it character now remain, though there is a nucleus on which historians, antiquarians, and others with a feeling for the past will build. The government of New South Wales has declared Hill End to be a historic village, and this declaration aims at guaranteeing its preservation and development as a national memorial. Already it is a vintage gold town.

Alluvial gold was found in the Hill End and Tambaroora areas during 1851, and three years later Tambaroora housed more people than any other New South Wales diggings. Twenty-six public houses thrived there, but the village no longer exists.

Hill End endured. In 1860, the rich Hawkins' Hill reef was opened, though ten years were to pass before it was fully exploited. By then, over 200 mining companies were operating. Some, such as Holtermann and Beyer's claim with its yield of 1,400 ounces from the first crushing of twenty-eight tons, gave brilliant returns. Others were the wildcat ventures of gamblers and rogues.

Hill End's peak period was 1872-1873. More than 30,000 people toiled, roistered, dreamed, or laboured in vain, but the atmosphere must have vibrated with golden excitement as 143,420 ounces were won. The village boasted a mile of shops, five banks, two newspapers, and twenty-eight hotels. One of these, the Royal, still serves the district after a century of continuous licence.

TUMBARUMBA The Black Tree

The story of Tumbarumba is short because it was a most difficult field to reach, and for the majority of diggers the results were not worth the effort. It became an alluvial mining centre in the late 1850s, and a jumping off place for the higher alpine diggings. The surrounding country is steep and thickly grassed.

Production was never large. In 1864, when gold was still pouring forth from other fields, Tumbarumba had only 1,533 ounces to be carried out by the gold escort. The village had a post office and police station by that time, and mounted mailmen carried the mail in and out twice each week. Oxide of tin, rubies, sapphires and zircon have also been found at Tumbarumba.

GRENFELL Derelict Pub

GRENFELL Old George Street

Probably more Australians know Grenfell as the birthplace of Henry Lawson than as the site of gold diggings of some note just over a century ago.

Cornelius O'Brien made the first gold strike, in dry, scrubby country on Emu Creek. When the mining registrar visited the area on 16 November 1866, he found that a rambling street had been formed and a store opened. Twenty-eight days later a name was found for the gold field and township when T. G. Grenfell, a district gold commissioner, was shot by bushrangers at Narromine, then a sheep property owned by William Charles Wentworth on the Macquarie River.

For two or three years, Grenfell was the richest field in New South Wales. Rich alluvial leads and quartz were worked, in depths from forty to eighty feet, and as much as twenty ounces to a load of dirt was washed in some of the claims. More than half of the field's total production of 350,000 ounces was won before 1871, after which it began to slow down until most of its diggers joined the 1874 rush to Parkes.

As on every gold field, some men flourished while others nearly starved. When Henry Lawson was born in a tent on 17 June 1867, his father was scratching a meagre living from the alluvial gravels. Perhaps Lawson was "imprinted," as ornithologists say of fledgling birds, with that consciousness of arduous and ill-paid labour which informed much of his work.

Festivals in honour of Henry Lawson have revived interest in the old town with its narrow, winding George Street and the obelisk which marks the tent site where he was born.

KIANDRA Decaying Shacks

KIANDRA Homestead Remnants

Kiandra was Australia's highest and least hospitable gold field, but for a few months it was one of the richest. The two brothers Pollock, who took sheep from the Murray to the Eucumbene district for summer grazing, found gold in November 1859. The find was reported to Tumbarumba, a rush began, and fortunes were made in the first weeks of 1860. On 15 February a gold commissioner reported that 1,500 were on the field, and that many claims were yielding up to a hundred dollars worth of gold per man every week.

The district was described as being about fifty miles north-east of Tumbarumba, on steep mountain ranges above the Tumut River. The diggings were considered to be habitable for only three months in a year, and in winter it was trackless through the deep snow drifts. Yet, by April 1860, 15,000 diggers had toiled into Kiandra.

A town quickly developed, mostly of tents on flimsy frames and primitive shacks of timber and bark. But it had hotels, one with a ballroom, piano, and a few female partners. The restaurant of Kidd's Hotel was the only place to buy a good hot meal served with knives and forks, and the Yankee proprietor built it long and narrow, with only one door in a corner, so that diners could not eat and run.

A newspaper was published from 3 August 1860 to 21 March 1861, when the proprietors wrote that the declining state of the diggings obliged them to "remove our plant to another place for the time being if not entirely."

Kiandra was short lived. In 1860, diggers won 67,687 ounces; in 1861, 16,565 ounces; in 1862, a little over 8,000 ounces; in 1870 a mere 1,600 ounces.

The Kiandra Snow Shoes Club was formed in 1878 or 1880; the first in Australia. Snowshoes had been used for transport and pleasure from Kiandra's earliest years. Today, the once glamorous alpine diggings stand at a strategic position on the junction of the Snowy Mountains highway and the access roads to the Snowy Mountain scheme. This, and the growing popularity of winter sports and tourism, has helped to revive the village.

OPHIR The Diggings

Ophir was never a large gold field, and did not produce riches similar to those of the fabulous oriental country after which it was named. Its main importance is historical, because it was the first gold field in Australia.

On 15 February 1851, Edward Hargraves and J. H. A. Lister found gold there when Hargraves panned the first gold from the alluvial gravels of Lewis Ponds Creek, a tributary of the Macquarie River. He proclaimed it to the world as the first payable gold field in Australia, and his discoveries were followed up by more gold found by Lister and William Tom, in April 1851. The Ophir gold field was proclaimed by the government in May 1851, but did not realise the hopes of its finders.

800 diggers had rushed the field by August 1851, but their earnings were never great. The first finds were limited and soon exhausted, and one estimate was that diggers never earnt more than twenty shillings a day.

The finding of Kerr's Hundredweight in the neighbourhood gave fresh impetus to their search. This gold mass weighing 106 pounds was found in a quartz outcrop by one of the Aboriginal station hands of Doctor Kerr, through whose property the creek ran. Kerr rewarded him with two saddle horses and full equipment, suits of clothing, blankets, and two flocks of sheep; mostly woolly sheep in lamb. Immediately there was killing, feasting, and dancing, and the remnants of the flocks were sold for a trifle. Doctor Kerr is said to have received £12,000 for his "hundredweight" from a bank in Bathurst.

Few signs of Ophir's brief glory remain. The ground is hungry-looking, the background trees are parched, and weeds smother mullock heaps and gouged-out gullies. An obelisk commemorates the finding of Australia's first payable gold.

NEW
SOU
Wyalong
Grenfell
Cowra
Young
Gundagai
Tumut
Adelong
Batlow
A.C.T.
Tumbarumba
Kiandra